Science of Electricity

Volume 8

Natural Gas
and Other Hydrocarbon Technologies
Explained Simply

by Mark Fennell
© 2012

This book is part of the
Energy Technologies Explained Simply™ Series

Other Books in the Energy Technology Series

<u>Renewable Energy Books</u>
Introduction to Electrical Power
Hydropower Technologies Explained Simply
Wind Power Technologies Explained Simply
Solar Power Technologies Explained Simply

<u>Coal Power Books</u>
Coal Power Technologies Explained Simply
Formation and Mining of Coal
Clean Coal Technologies
Mercury and Coal Power

<u>Nuclear Power Books</u>
Nuclear Power Meltdowns and Explosions
Health Hazards of Radioactive Decay
Radiation Measurements
Processes of Radioactive Decay and Storage of Nuclear Waste

<u>Natural Gas Books</u>
Natural Gas Basics
Extracting and Refining Natural Gas (includes Fracking)
Transportation, Storage, and Use of Natural Gas

<u>Power Line and Grid Books</u>
Introduction to the Transmission of Electrical Power
Power Lines
Underground Cables
Utility Operations and Quality Control
Power Grids Explained Simply

About the Book

Introduction

This book discusses everything you need to know about Natural Gas. This includes extraction, refining, transportation, and safety. We will also discuss additional hydrocarbon fuels which are often used to produce electrical power, particularly biodiesel and biomass.

The information in this book will answer all questions you have regarding natural gas, including:

- Where can natural gas be found?
- How do we extract and refine natural gas?
- What are the environmental effects of fracking?
- How does the pipeline system work?
- What is the odor associated with natural gas?
- How can we minimize flammability and explosions?

This book will also discuss the uses of other types of hydrocarbon fuels including diesel, biodiesel, and biomass.

Chapter Topics

Chapter one provides an overview of hydrocarbon fuels. In this chapter you will learn how hydrocarbon fuels are formed. You will also learn about the different types of hydrocarbons which can be used as fuel.

Chapter two provides an overview of Natural Gas. In this chapter you will learn about the chemicals associated with natural gas and the energy which can be obtained from natural gas. You will learn the advantages and disadvantages of natural gas. Significant time is devoted to the hazards of natural gas, particularly hydrogen sulfide and explosions.

Chapter three discusses how we can find and extract natural gas. Natural gas can be found in a variety of locations. However, extracting natural gas requires us to consider factors which are unique to each location. The options will be discussed here.

Chapter four discusses fracking. Fracking is a general term for using a high pressure mixture to break apart the rocks which trap natural gas. Fracking is commonly used by industry, yet can severely harm the local environment. Many citizens are understandably concerned. Therefore in this chapter we discuss fracking in great detail, including the process, the environmental impact, and the possible options.

Chapter five discusses the refining operation for natural gas. Raw natural gas contains numerous chemicals other than the natural gas, and therefore must be refined significantly before the gas can be used. This chapter explains all of the major processes for refining natural gas.

Chapter six discusses the transportation and storage of natural gas. The primary transportation system for natural gas is the pipeline system spread throughout the nation. Therefore this chapter discusses the three main pipeline systems: the Gathering System, the Transmission System, and the Distribution System. We begin with a general overview of the pipeline systems. This is followed by discussions of pipeline construction, pipeline corrosion, pipeline inspection, and natural gas compressors.

In chapter six we also discuss an alternate method of transporting natural gas: liquefied natural gas. We discuss the process of compressing gas into liquefied natural gas. This is followed by practical tips regarding transportation and storage of the liquefied natural gas.

Chapter seven takes us in a new direction, discussing a variety of other hydrocarbon fuels which are used for the creation of electrical power. Topics include diesel, biodiesel, biomass, and cogeneration.

In the Appendix you will find brief explanations of terms and abbreviations commonly used by the Natural Gas Industry.

The book is completed with a comprehensive index which will help the reader find any topic easily.

About the Energy Technology Series

Purpose of this Series

The books in the *Energy Technologies* series are designed to educate citizens, students, and legislators on all aspects of energy technologies. The first books in the series focus on electrical power.

The books discuss many energy technologies, including: generators, turbines, power plants, power lines, and grids. The technologies for each type of power source (hydro, wind, solar, coal, nuclear, and natural gas) are discussed in detail. The books also discuss efficiency, safety, reliability, and health concerns for each energy technology.

The ultimate goal of the series is to enable the people to make informed decisions on practical energy questions. The secondary goal is to serve as introductory guides for students embarking on careers with energy technologies.

Taken altogether the books in the series answer any question you are likely to have, such as:

- How can we increase the efficiency of solar cells?
- How do I select the size my solar array?
- What do I need to know when installing a wind turbine?
- How effective are the clean coal technologies?
- How can we prevent grid failures?
- Do power lines cause cancer?
- and many other energy technology questions…

Science of Electricity in Perspective

The subject of electrical power is of great importance to our communities, but is rarely taught. Public debate is frequent and passionate, but with too little understanding of the actual science. At best, an informed citizen knows only a few pieces. At worst, as it is for a great number of citizens, electricity is magic and myths are believed as scientific truth. It does not have to be that way. Any citizen, regardless of background, can know the technologies behind all aspects of electricity.

The books in this series solve that problem. These books educate the general public in all aspects of electrical power. Any person, regardless of background, can easily find the answer to his energy question in one of these books.

Specific Goals

There are numerous technologies described in these books. Yet for each technology I sought out the answers to the following questions:

1. How does the technology work?
2. What are the advantages and disadvantages?
3. What is the efficiency? How can the efficiency be improved?
4. What is the environmental impact? How can it be improved?
5. What are the safety hazards, and how can they be reduced?
6. What are the most important practical tips?
7. What facts comprise the most important data?

Technical Discussions Explained Simply

The books in this series explain the principles of electricity as simply as possible, using ordinary English (no engineering jargon), and highlighting the most important points of each technology. Main concepts and facts are emphasized with the use of lists, tables, diagrams, and summaries.

I do not expect any reader to have a background in science, yet I offer enough facts and details so that the reader can have an accurate understanding of all related technologies. I provide enough technical details and enough data for the reader to make informed decisions.

Conclusion

For all the reasons above, I offer this series of books. My goal is to inform you on the basic concepts of all the technologies and all of the issues related to electricity so that you can make realistic decisions.

Remember that there are no perfect solutions, there are only choices. I hope that this series of books will assist you in making those choices for your community.

Mark Fennell

Accuracy and Technical Depth

Objectivity

I have tried my best to be as objective as possible. Whereas many other authors of energy books have an agenda, I have no desire to promote one industry over another. I have no desire to promote one technical solution over another. In this endeavor, I have tried to be an objective scientist.

Accuracy of Data and Summaries

I never relied solely on the conclusions of other researchers. Instead, I performed many other tasks to ensure that all conclusions were accurate. I examined primary data whenever possible. I have read the fine print on how research was obtained.

I have also checked the accuracy of the conclusions written by other researchers, most commonly by finding at least three distinct sources for each fact. In addition, I performed my own calculations numerous times to prove (or disprove) conclusions and final values in other reports. It is only after such rigorous investigations that I created data tables and wrote summaries for these books.

Limited Mathematics

The books must also use math from time to time. However, I want to emphasize that I focus on concepts not on the mathematics. I provide equations only when it is necessary for the citizen or student to be familiar with these equations.

M.F.

Table of Contents

8.1
Overview of Hydrocarbon Fuels

Introduction

Most of the other fuels used for electric power fall into the category of hydrocarbon fuels. Some of these fuels are commonly known as fossil fuels. However the term "fossil fuels" indicates the origin and location, not the chemical composition. Many of these same fuels can be obtained from other methods and other locations.

The most important hydrocarbons for creating electricity are natural gas, diesel fuel, and biomass. Each of these will be discussed in more detail in separate chapters.

Hydrocarbons are carbon chains with hydrogen atoms attached. When hydrocarbons are burned, heat is released. The by-products are CO_2 and H_2O, both of which are safe to humans and to the environment.

The number of carbons in a hydrocarbon can vary. Note that more carbons in a hydrocarbon molecule will result in more energy being released per molecule when the fuel is burned. This is an important concept when selecting fuels.

Hydrocarbons, Plant Life, and Fossil Fuels

Plants and animals, at their most basic, are complex carbon chains. When plants and animals die their carbon chains can be used as fuel for burning. Plants and animals which are no longer living are referred to as "organic material." If the organic material was created recently, then it is called "biomass." The most common forms of biomass are firewood and agricultural waste.

If the organic material has been within the earth for thousands of years then it is called a "fossil fuel." Fossil fuels include natural gas, gasoline, oil, and coal. The formation of fossil fuels involves temperature, pressure, and time. Essentially, dead plants and animals are buried under layers of soil. The high pressures on the organic matter, the high temperatures of the region within the earth, combined with particular bacteria, will convert the organic material into hydrocarbon fuels.

Hydrocarbon fuels range from natural gas (the simplest fuel), to coal (the longest chain of any hydrocarbon fuel), with hundreds of variations in between. The types of hydrocarbon fuels created depend on several factors, including: the types of organic material, types of bacteria, pH level, pressure, temperature, and length of time.

Types of Hydrocarbons

Hydrocarbons are specified foremost by the number of carbons in the chain. The hydrocarbons most commonly used for fuels have between 1 and 40 carbons in the chain, although coal has far more. In total, there are hundreds of possible hydrocarbons.

Types of hydrocarbons are often grouped together in categories. These groupings are usually arranged according to any one of the following factors: energy value, number of carbons, phase at room temperature, boiling point, or common use.

Table 8.2: Grouping Names and Number of Carbons

Grouping Name(s)	Number of Carbons	Phase
Natural Gas	1 Carbon	Gas
Light Hydrocarbons	2-4 Carbons	Gas
Napthas, Natural Gas Liquids, Condensate	5-7 Carbons	Liquid
Gasoline	7-12 Carbons	Liquid
Fuel Oil (includes kerosene, diesel fuel)	10-40 Carbons	Liquid
Kerosene	10-15 Carbons	Liquid
Diesel fuel	10-20 Carbons	Liquid
Lubricating Oils	16-20 Carbons	Liquid
Tar, Asphalt	40-70 Carbons	Solid

Other Molecules Associated with Hydrocarbons

Introduction

There are several other molecules commonly associated with hydrocarbons, particularly with fossil fuels. The most common of these other molecules include sulfur compounds and water. Other molecules may include nitrogen compounds, carbon dioxide, and various minerals.

Sulfur Compounds

Most hydrocarbons naturally exist with sulfur compounds. The first problem with sulfur, particularly hydrogen sulfide, is the toxicity. Hydrogen sulfide is highly toxic and is usually in the gas phase. Therefore, a person can easily inhale hydrogen sulfide and be seriously harmed by it.

The second problem with sulfur, in any form, is that burning of sulfur creates SOx molecules. These molecules contribute to acid rain. Therefore sulfur must be removed from the hydrocarbons whenever possible.

Sulfur compounds as found with hydrocarbons usually exist in one of three forms. In the gas phase, sulfur usually exists as hydrogen sulfide. In the liquid phase, sulfur usually exists as a thiol. In the solid form, sulfur usually exists as a pyrite.

The most common of these sulfur compounds is hydrogen sulfide, H_2S. Hydrogen sulfide exists as a gas and is highly toxic.

Thiols are similar in chemical structure to alcohols, but with a sulfur atom in place of the oxygen atom. Thiols are colorless, but they have a slight odor. Thiols are usually found as liquids, but can be in the gas phase.

Pyrites are various solids chemically made of iron sulfides. Pyrites come in several shapes and colors. Pyrites can be formed by several processes; one common process is the reaction of hydrogen sulfide with iron.

Water

Fossil fuels usually have water. There are numerous biological and geological causes for water being with the fossil fuels (reasons are beyond the scope of this book). The basic problems with water are related to the transporting of hydrocarbons. Water adds weight and increases volume. Water can rust containers. Water can also impede the flow of hydrocarbons through various pipes. Therefore, water must be removed whenever possible before the hydrocarbons are transported.

Summary of Hydrocarbon Fuels

1. Hydrocarbons are carbon chains with hydrogen atoms attached. When hydrocarbons are burned, heat is released. The by-products are CO_2 and H_2O.

2. More carbons in a hydrocarbon will result in more energy being released per molecule.

3. Many of hydrocarbon fuels are commonly known as fossil fuels. However, the term "fossil fuels" indicates the origin and location, not the chemical composition.

4. Hydrocarbons are specified foremost by the number of carbons in the chain. The most commonly used hydrocarbons have between 1 Carbon and 40 Carbons in the chain.

5. Because there are so many hydrocarbons these hydrocarbons are often grouped together. Groupings are usually arranged by number of carbons, phase, energy value, or common use.

6. There are other molecules commonly associated with hydrocarbons. The most common of these other molecules include sulfur compounds and water. Other molecules may include carbon dioxide, nitrogen, various minerals, and trace amounts of other gasses.

7. Hydrogen sulfide is highly toxic. It exists in the gas phase, often in high concentrations, and with most sources of hydrocarbons. Therefore H_2S is a common health hazard when extracting hydrocarbon fuels.

8. The basic problems with water are related to the transporting of hydrocarbons. Therefore water must be removed whenever possible before the hydrocarbons are transported.

8.2
Overview of Natural Gas

Introduction

Natural Gas is Pure Methane

Natural gas is pure methane. It is the simplest of hydrocarbons. Natural gas is a very clean fuel. Natural gas can be purified to nearly 100% pure methane. When burned, natural gas creates only carbon dioxide and water.

When natural gas is extracted from the ground there may be several other molecules with the methane. These other molecules are separated from the raw supply and only pure methane is sent to be used.

Methane is colorless and odorless, therefore methane can be difficult to detect. For safety reasons, distribution companies are required by law to put a distinct odor in the gas. The odor added is usually a form of sulfur, specifically a type of thiol. If you "smell" natural gas, you are actually smelling the added odor of the sulfur compounds, not the gas itself.

Formation of Natural Gas

Natural gas (methane) is formed in many ways. The most common method is the decay of organic material. Traditionally, most methane has been obtained from wells such as petroleum wells. The formation of natural gas at these locations is the same process as the formation of any fossil fuel. However, because methane is the simplest hydrocarbon we do not necessarily have to wait thousands of years. We can also obtain methane from species of plants which have died much more recently.

Energy of Natural Gas

Natural gas is measured by volume, usually by the cubic foot. One cubic foot of natural gas has approximately 1,027 BTUs of energy. Comparing natural gas to coal we can state that approximately 25,000 cubic feet of natural gas has the equivalent energy of one ton of coal.

<u>Supply of Natural Gas</u>

Estimating the supply of natural gas is very difficult. The experts at various natural gas organizations come up with different values. The most conservative of estimates of natural gas in the United States is approximately 1,000 Trillion Cubic Feet. The general consensus is that we enough natural gas to last us many years. Most experts have concluded that there is enough natural gas for companies to invest in natural gas power plants and operate these power plants for a significant length of time.

Advantages and Disadvantages of Natural Gas

There are several advantages to using natural gas for generating electricity:

- Natural gas can be stored.
- Natural gas burns efficiently.
- Natural gas generators are quick to start up.
- Main pipelines for natural gas exist in many areas already.

There also several disadvantages of using natural gas for generating electricity:

- Natural gas is a non-renewable resource.
- Natural gas often requires sophisticated refining.
- Natural gas requires an extensive system of pipelines.
- Natural gas is flammable, often explosive.

Using Natural Gas for Electrical Power

Introduction and Major Steps

Natural gas is most commonly used (and most effectively used) for heating and cooking, yet gas is also being used to create electricity. In order to use natural gas for electrical power we must go through several important steps:

1. Find and extract raw natural gas
2. Refine natural gas (into pure methane)
3. Transport natural gas through pipelines
4. Store natural gas
5. Use natural gas as fuel for turbines

Each of these steps will be discussed in subsequent chapters.

Turbines for Natural Gas

When natural gas is used to create electricity, the gas is burned and the resulting energy is applied to the turbine. Natural gas can be used with a gas turbine, a steam turbine, or a combined cycle turbine.

The process for creating electricity from natural gas is very similar to the process for coal. When natural gas is burned, heat is released. The heat is used to boil water. The water becomes steam, the steam pushes the turbine, and the turbine operates the electrical generator.

Like coal power plants, natural gas power plants can use combustion turbines, steam turbines, or both turbines in succession.

Best Use

Electrical generators which use natural gas are best suited as Peak Use generators. Reasons for this include:

1. The process can be started and stopped very quickly.
2. The natural gas can be stored.
3. The process is very efficient.

Hazards of Natural Gas

Introduction

Although natural gas is has many benefits, natural gas also has several hazards. The most significant hazards include:

1. Raw natural gas often contains H_2S, which is very toxic.

2. Raw natural gas contains other chemicals, which can be flammable, toxic, or both.

3. Methane is very flammable and can be explosive in closed areas.

Hydrogen Sulfide (H_2S)

Hydrogen sulfide is very toxic and exists in the gas phase. Furthermore, natural gas wells contain high concentrations of hydrogen sulfide. Therefore any leak in the collection system will be very harmful to a person working nearby. Hydrogen sulfide has been known to cause permanent damage and death to workers in the natural gas industry.

Toxic Molecules in Raw Natural Gas

Within a mixture of raw natural gas, there are many chemicals which are harmful if inhaled or swallowed. Some of these are carcinogens. Some will harm the environment.

Note that we cannot specify the molecules (beyond H_2S, which is always there) because each source will have different chemicals, and these chemicals will exist in varying amounts. Therefore each natural gas source must be analyzed for specific chemicals. Steps to remove some of these molecules will be discussed in later chapters.

Flammability of Natural Gas

Purified natural gas (methane) is highly flammable. Remember the entire purpose of using natural gas is to burn it as fuel. Therefore if conditions are right then the methane can burn in any pipeline or container.

Methane can also be quite explosive. If the concentration of methane is high enough then the methane can explode simply by the trigger of a small spark or the heat coming from any machine nearby.

Raw natural gas is even more flammable. Raw natural gas and crude oil may contain dozens of different types of hydrocarbons, many of which are suitable for fuel. Some of these hydrocarbon fuels include: methane, propane, gasoline, and diesel gas. Therefore this mixture can and will burn given the right conditions.

However there is a distinct difference in the fire hazards of the methane which exists outdoors versus the methane which exists indoors. The methane which exists outdoors is usually not a fire hazard, but methane indoors can be a serious fire hazard. These will be discussed below.

Fire Hazard of Methane Outdoors

Methane outdoors is not much of a fire hazard. Methane is very light. It will naturally rise through the ground and up into the air. If there is a leak in the underground pipe then there will not be much methane above ground to burn. Most of the methane will be in the air.

Also, because there will not be much methane above ground outdoors, when a fire in the area does exist (such as a brushfire), there will probably not be a methane fire.

Fire Hazard of Methane Indoors

The situation is very different when methane exists indoors. When methane is contained indoors, the potential for fire is great.

The final pipelines for methane will always lead the methane indoors. If there is a leak in these pipes then the methane will rise into the room. Because the room is a closed area, the methane will *not* go away. The methane will continue to build. At this point all that is needed to start a fire is a small spark or sufficient heat. A hot day, a warm room, or the ignition of a stove...any of these sources of heat will be sufficient to start the methane burning.

Also note that a greater amount of methane built up in the room will result in a larger fire, often resulting in explosions. When the amount of methane reaches 5% of the total air in the room, then an explosion is likely to occur.

Minimizing Methane Explosions

There are several methods which have been developed to minimize methane explosions:

- Add an odor to the methane, so that leaks can be smelled quickly.
- Install methane detectors, which automatically sample the methane in the air and give a digital display of methane concentration.
- Check all pipes and containers for leaks annually. Check more often where flames are used nearby.

Summary of Natural Gas Overview

1. Natural gas is methane, CH_4.

2. When natural gas is extracted from the ground there may be several other hydrocarbons. These other molecules must be separated from the methane.

3. Natural gas is formed by organic material which has decayed over many years. The formation of natural gas is similar to the formation of coal and oil.

4. Common locations to find natural gas include: coal mines, oil wells, natural gas wells, and landfills.

5. Estimating the supply of natural gas is very difficult. The general consensus is that we enough natural gas to last us many years.

6. Natural gas can be used to create electricity. This is done by burning the gas, then applying the energy to a turbine and generator.

7. Natural gas power plants can use combustion turbines, steam turbines, or both turbines in succession.

8. Electrical generators which use natural gas are best suited as Peak Use generators:
    a. The process can be started and stopped very quickly.
    b. The natural gas can be stored.
    c. The process is very efficient.

9. In order to use natural gas for electrical power, we must go through the following steps:
    a. Find and extract raw natural gas
    b. Refine natural gas (into pure methane)
    c. Transport natural gas through pipelines
    d. Store natural gas
    e. Use natural gas as fuel for turbines

10. Advantages of Natural Gas
    a. Natural gas can be stored.
    b. Natural gas burns efficiently.
    c. Natural gas generators are quick to start up.
    d. Main pipelines for natural gas already exist in many areas.

11. Disadvantages of using Natural Gas
    a. Natural gas is very flammable.
    b. Natural gas is a non-renewable resource.
    c. Underground pipes adds logistical problems and safety concerns.
    d. Raw natural gas is often toxic, flammable, or both.

12. Natural gas has several hazards, including:
    a. Raw natural gas often contains H_2S, which is very toxic.
    b. Methane is very flammable.
    c. Raw natural gas contains chemicals which can be flammable, toxic, or both.

13. Methane which exists outdoors is usually not a fire hazard. However, when methane is contained indoors the potential for fire is great.

14. There are several methods which have been developed to minimize methane explosions:
    a. Add an odor to the methane so that leaks can be smelled quickly.
    b. Use methane detectors which automatically samples the methane in the air and give a digital display of concentration.
    c. Check all pipes and containers for leaks frequently.

8.3
Finding and Extracting Raw Natural Gas

Introduction

There are two general approaches to finding natural gas:

1. Look for reservoirs of natural gas which were made thousands of years ago

2. Capture natural gas as it is being created today from recently decaying organic material.

Methane is most commonly found in the following types of locations: coal mines, oil wells, natural gas wells, landfills (recently decaying organic matter), and swamps.

The techniques for finding methane depend on the type of area we are searching. The techniques for extracting methane also depend on the specific molecules which co-exist with the methane at that location. In this chapter we will discuss best practices for finding and extracting natural gas in each type of location.

Finding and Extracting Natural Gas from Coal Mines

Remember what we said in the chapters on coal: wherever there is coal, there will be methane.

Extraction of methane from coal mines can either be passive or active. Passive extraction relies solely on the nature of the gas. Active extraction uses an external power source to pull the methane out of the mine. In passive extraction, we insert a tube into the mine. Methane gas naturally rises so it travels through the tube and into the collection containers. In active extraction a power source is used to draw methane through the tube.

Methane removal systems (these are not extraction systems) already exist in many coal mines. These exist because the methane must be removed from the coal mine to prevent explosions. However, few of these systems actually capture the methane. In most cases, coal is the fuel of interest and the methane is simply let out into the air. With a few modifications many of these systems can be transformed into capturing methane efficiently.

Finding and Extracting Natural Gas from Oil Wells

The most common source of natural gas is the oil well. A reservoir of oil often contains many hydrocarbons, including methane. The methane is often dissolved in the crude oil. The methane can also be "free" (not dissolved in the crude oil) yet trapped in the same reservoir where the oil is located.

If the location is considered an oil well then the energy company focuses on extracting oil, not natural gas, and uses standard oil pumping techniques to collect the oil. However, these pumping techniques will collect all hydrocarbons in the well, including methane. Each hydrocarbon fuel is then separated at the refinery. Methane can be collected in this way.

Note that the presence of methane in an oil well assists the pumping operation. Methane is a light gas which naturally floats to the air. When this light gas is dissolved in the oil, the methane helps lift the oil to the surface. Furthermore, a higher concentration of methane in the oil will result in less pumping required. Of course, the methane must be separated from the mixture to be useful as natural gas.

Sometimes methane is trapped in pockets behind the oil well. In this case, the well is first pumped for oil and other liquid hydrocarbons. Then the company uses fracking techniques to extract the methane. (See the chapter on Fracking for more details).

Finding and Extracting Natural Gas from Natural Gas Wells

There are many locations where methane exists by itself. This means that no oil or coal was created in the same region. Typically, the deeper into the earth we go, the reservoir will be more methane and less oil.

Finding such deep reservoirs is difficult to do from the earth's surface. However, technological advancements in the past decade have improved the odds of finding methane reservoirs. Sonar methods are the most effective. Engineers also rely on knowledge of the geology acquired from surveys in nearby regions.

If the reservoir is almost entirely methane then the methane can be extracted passively. A well is created, the pipe is inserted, and the methane will naturally rise up to the surface.

This natural gas well may contain other light hydrocarbons, such as ethane and propane. Some pumping may be required, depending on the other hydrocarbons. These hydrocarbons must also be separated into different containers.

Note that a reservoir might seem empty of methane yet not really be empty. Often there is more methane trapped inside porous rocks. There are several ways to access this methane, the simplest of which is using acid. Acid will dissolve the rocks, thereby freeing the methane. Also note that this technique is a major component of the fracking process (which will be discussed in detail later).

Finding and Extracting Natural Gas from Landfills

Up to this point we have talked about plants which decayed millions of years ago. Yet methane is also created by plants which have decayed recently.

Methane is the simplest hydrocarbon. Therefore, we can always count on methane being produced from decaying matter. Other hydrocarbons might or might not form, but we can be fairly certain that methane will always be created.

The greatest collection of recently decaying plant matter can be found in landfills. The gas being emitted from a landfill is approximately 50% methane and approximately 45% carbon dioxide. The remaining 5% of molecules in the gas phase are primarily nitrogen, oxygen, and hydrogen sulfide. (See data table)

Molecules in Raw Landfill Gas

Molecule	Range	Most Common
Methane (CH_4)	35%–60%	50%
Carbon Dioxide (CO_2)	35%–55%	45%
Nitrogen (N_2)	0%–20%	5%
Oxygen (O_2)	0%–2.5%	<1%
Hydrogen Sulfide (H_2S)	no data	<1%
Water Vapor (H_2O)	1%–10%	no data
NMOCs*	<1%	<1%

*NMOC = Non-Methane Organic Compounds
**Data is taken directly from Energy Information Administration

Extracting methane in a landfill can range from the simple to the sophisticated. The simplest method is to insert a tube into the ground. As methane is created it will naturally rise. This gas travels through the tube and is collected in a container.

The next stage in sophistication is to use external power to actively suck up the gasses. In this method, some type of motor is connected to the tubes and the gas is pulled up from the ground.

In the most sophisticated methods of extraction, gas collection tubes are placed along the ground, then the trash and layers of soil are put on top. As the trash decays, methane and carbon dioxide are created. These gasses enter the underground tubes, which then carry the gasses to a collection area. This method can be passive, but is usually active (where an external source pumps the gasses through the tubes).

Note that the first two methods can be done after the landfill has been created, but the third method requires that the equipment be installed before adding the trash. Note also that there is no reservoir. We are collecting the methane as nature is creating it.

Finding and Extracting Natural Gas from Swamps

People familiar with swamps have long known that swamps are a source of methane. For our purposes, swamps are essentially traps for organic material. Many plants and animals are captured by the swamp. This organic material often drops to the bottom. There, at the bottom of the swamp, the organic material decays. Because of the lack of oxygen down there, many anaerobic bacteria thrive on this material, converting this material into various molecules. The net result is the creation of many gasses, one of which is methane.

Note that swamps usually emit several gasses. Common gasses emitted from a swamp include: methane, carbon dioxide, nitrogen, and hydrogen sulfide. Swamps also emit some VOCs. (See VOCs in the section below). Other hydrocarbons are often formed, though usually in less quantities. Some of these are gasses, such as ethane. Other hydrocarbons are liquids such as pentane and hexane. In some cases, several heavier hydrocarbons (collectively grouped as "crude oil") can be found in a swamp.

Although people have known about swamps as a source of natural gas for a long time, swamps have not been a major source of methane collection. The large amount of water makes logistical problems, which have historically been a deterrent. The theoretical options for extracting methane include:

- pumping the swamp, then separating the water from the gas.
- laying a pipe at the bottom of the swamp, collecting both water and gasses, then separating the gas from the water.
- building a trap above sections of the swamp, which will capture gas as the gas rises.

Volatile Organic Compounds (VOC)

Landfills and swamps emit gasses which are known as Volatile Organic Compounds (VOC). Note that the term "Volatile Organic Compound" is very generic. There are hundreds of molecules which can be classified as a VOC. Some are safe, some are harmful. The only facts which can be stated for certain about VOCs are: the molecules contain carbons, and the molecules are in the gas phase at room temperature.

Collectively, VOCs are less than 1% of the molecules formed in a landfill, but some of them can be toxic. Note that some of these VOCs are hydrocarbons other than methane (such as ethane and propane).

Another term often used for these molecules is "Non-Methane Organic Compounds" aka NMOCs.

Flaring vs. Collection

In many landfills the methane is extracted for safety reasons but is then burned away. Burning the methane is known as "flaring." Note that environmental regulations mandate that landfills control the amount of methane created, but there are no regulations regarding methane collection.

Environmental regulations and safety regulations mandate that landfills control the amount methane. Because of this there are many landfill companies which install underground pipes, along with a vacuum, in order to remove the methane. At this point there are two options: burn it or collect it.

Many landfill companies find it simpler to burn the methane rather than collect it. The main reason a landfill company chooses to flare rather than collect the methane is the business perspective.

Many of these companies see themselves as a waste management company, not an energy company. The company is not interested in building a refinery, nor is the company interested in building a pipeline system to energy customers. Therefore the methane is burned rather than collected.

Instead of flaring, there are ways the methane can be collected and distributed profitably. The methane produced by a landfill can be sold to an energy company. Alternately, a waste management company can itself become an energy company. Both of these methods have been used successfully by waste management companies in the United States.

Note also that many oil companies burn their methane rather than collecting it. They see gasoline, diesel, and other liquid hydrocarbon fuels as their main objective. Methane is considered not as profitable, and not worth keeping. Therefore they simply burn the methane rather than collect it as another useful fuel.

Some other energy companies burn methane to prevent explosions. Methane is more explosive and quicker to burn than gasoline, yet combined with the energy in gasoline a small explosion can quickly turn into a large fire. For this reason, some oil companies choose the flaring of methane rather than collecting it simply as a matter of safety.

Regardless of the reasons, the result is the same: a non-renewable useful fuel is lost, without any useful application of that energy.

Summary of Extracting Natural Gas

1. In order to use natural gas for electrical power we must go through the following steps:
   a. Find and extract raw natural gas
   b. Refine natural gas (into pure methane)
   c. Transport natural gas
   d. Use natural gas as fuel for turbines

2. Pure natural gas is only methane. Methane is the simplest of hydrocarbons. Therefore methane forms easily.

3. Raw natural gas usually contains other molecules and therefore raw natural gas must be refined before using.

4. Methane is most commonly found in the following types of locations: coal mines, crude oil reservoirs, natural gas reservoirs, landfills, and swamps.

5. Extraction methods can be categorized as passive or active. Passive extraction relies solely on the nature of the gas: the gas naturally rises, traveling through tubes and into containers. Active extraction uses an external power source to pull the methane out of the reservoir.

6. Methane from Coal Mines: Wherever there is coal, there will be methane. Because the methane must be removed from the coal mine in order to prevent explosions, methane removal systems already exist. However, many of these systems simply vent the methane rather than collect it.

7. Methane from Crude Oil Reservoirs: Where there is crude oil, there will usually be methane. The crude oil reservoir is the most common source of natural gas simply because of the economics and infrastructure of crude oil. Extraction can be passive or active, but usually active extraction is used. Separating methane from crude oil requires significant refining.

8. Methane from Natural Gas Reservoirs: Natural gas can be found by itself in natural gas reservoirs. Finding these reservoirs can be difficult because they usually exist deep within the earth. Much of the methane will be trapped inside porous rocks. Extraction can be passive or active.

9. Methane from Landfills: Landfills produce methane regularly. The gas being emitted from a landfill is approximately 50% methane and approximately 45% carbon dioxide. Extracting methane in a landfill can range from the simple to the sophisticated. The gas from landfills can easily be collected. However in many landfills the methane is burned away ("flared") rather than collected. Flaring is done more for business reasons than science reasons.

10. Methane from Swamps: Swamps usually emit several gasses, including methane, carbon dioxide, nitrogen, and hydrogen sulfide. Other hydrocarbons may be formed in swamps, though usually in less quantities. Swamps also emit some Volatile Organic Compounds (VOCs). Swamps have not been a major source of methane collection because the large amount of water makes logistical problems.

11. Landfills and swamps emit gasses which are known as Volatile Organic Compounds (VOC). Collectively, VOCs are less than 1% of the molecules formed, but some of them can be toxic. Another term often used for these molecules is Non-Methane Organic Compounds (NMOCs).

8.4
Fracking for Natural Gas

Introduction

Fracking is a generic term for a variety of methods for obtaining natural gas from difficult to reach locations.

Most natural gas is contained in underground wells. These wells are essentially caverns. All you have to do is dig deep enough to reach the cavern, and the natural gas will be there. Once the well is found, the natural gas can be pumped out.

Yet there are many areas underground where natural gas exists, but is not in an easy to access cavern. The gas exists in small pockets throughout the region rather than as one large cavern. The gas can also be trapped within the porous rocks.

If the rock is relatively porous then the natural gas can flow in and out very easily. However, if the rock is not porous then the gas may be trapped behind the non-porous rock. The gas may also be contained tightly within that rock. Therefore simply drilling a hole is not enough, we must also break apart the rocks before we can access the natural gas. This is the purpose of fracking.

Therefore, fracking can be defined as any method which dissolves rocks or breaks rocks apart deep below the earth's surface so that small pockets of natural gas can be allowed to escape.

There are four main methods of fracking:
1. Water: Using water to gradually erode the rocks.
2. Water Pressure: Using an intense pressure of water to blast apart rocks.
3. Chemicals: Using chemicals to dissolve the rocks.
4. Sanding: Using a type of sand blasting to break apart the rocks.

In practice, all four items are usually used at the same time.

Fracking Process

The process of fracking involves sending a liquid under high pressures to dissolve the rocks.

First a hole is dug several thousand feet below the surface. Then a standard vertical pipe is installed. At the end of this pipe is a base. Attached to this base are a series of horizontal pipes. These horizontal pipes will spread the dissolving liquid to the rocks in a 180 degree or 360 degree direction.

Before the dissolving slurry is pumped in, the area above the base is often sealed. This prevents slurry chemicals from reaching the water tables and aquifers above the reservoir. This seal also helps contain the natural gas, making it easier to collect.

The water is pumped in at very high pressures to physically break apart the rock. Acid is added to the water, which dissolves the rock. Sand acts as a sandblaster, eroding the rock with abrasion. Sand is also used to prop open some of the holes. A small percentage of additives are also mixed in to help minimize pipe corrosion and to improve flow. The combination of high pressure water, sand blasting, and chemical reaction will eventually dissolve the rocks in the area.

The primary chemical additive is a type of acid. The acid reacts with the carbonates in the rock, turning the rock into water and carbon dioxide. A hole is created where the natural gas can escape. The dissolved rock also releases various other solid compounds.

Using this slurry of water, sand, and additives, at high pressure, will eventually dissolve the rock. After the rock has dissolved, the gas escapes. The natural gas is pumped out and collected.

If necessary, a subsequent fracking operation can be done, sending in more water with chemical additives, dissolving the rock, pumping the water out, and then collecting more natural gas. This can continue as long as the drilling company believes the operation is profitable.

Pumping out Waste Water and Natural Gas

After the dissolving liquid has performed its job, and before the natural gas is collected in large quantities, the liquid is pumped out as waste water. This waste water is a complex mixture. It may contain natural gas, water, sand, dissolved rock, and chemical additives. The chemical additives and the compounds from the dissolved rock may provide environmental hazards.

Note that some natural gas is also pumped out with the waste water, and therefore the natural gas must be separated. Over time, the solution being pumping out will be less water and more natural gas. Most of the waste water is pumped out within a few days (along with natural gas). Within a few weeks all the water is pumped out, leaving only natural gas to be pumped.

After the waste water has been removed (along with an initial supply of natural gas), the full supply of natural gas (originally trapped in the rock) begins to emerge. This natural gas is then pumped out and collected.

The waste water can be stored, reused, or filtered further. Note that the biggest concern of fracking is the possibility of contaminating drinking water with toxic chemicals from unfiltered waste water.

Environmental Hazards of Fracking: Overview

There are four primary environmental hazards from fracking operations:

1. The enormous amounts of water used, which compete with other water needs of the region.

2. Many of the chemicals used in fracking are toxic, and the chemicals used are in great quantities, which can significantly harm humans.

3. The waste water is filled with soil and chemicals, and therefore cannot be used for most other water uses.

4. The drilling operation is below levels of aquifers and underground water streams, and therefore the toxic chemicals from fracking can easily contaminate underground water supplies.

Water Requirements for Fracking

The water requirements for fracking are significant. A typical natural gas well will require between 2 million and 4 million gallons of water to do the job.

This water is essentially a non-renewable resource. The water contains so much sand, rock, soil, and chemical additives as to be useless for many other purposes. However, engineers have been developing technologies and methods to improve the situation. Some of these methods will be discussed below.

Where does the gas company get such large amounts of water? The water usually comes from local lakes, local rivers, or from underground water systems. The water can also come from the municipal water supply.

When fracking operations are in full production, the depletion of water is significant. The drilling company can be the largest consumer of water in the entire region. Fracking operations are a major competitor for regional water use, and if they are allowed to use the water then the citizens may be deprived of water for essential needs.

The best way to offset this use of water is for the drilling company to use its own water supply. Their water can be filtered and reused for each subsequent fracking operation. The company can also create its own supply of water from rainfall to partially offset the community water supply.

At the very least, the water management authorities in the region must carefully designate when the fracking operations can take place and how much water the drilling company may use at any given time. For example, some water management authorities have limited fracking operations to certain seasons when water is more abundant. Other authorities place priorities for water use, such as: drinking, agriculture, and recreation, only allowing water for fracking after those other needs have been met.

Chemicals Used

Overview

The biggest concern of fracking is the use of chemical additives. Some of these chemical additives are toxic to humans or are harmful to the environment.

How toxic are these chemicals? It is very difficult to say. The types of chemicals used are almost unlimited, and the amounts of these chemicals will vary from location to location. Each drilling company uses its own set of chemicals. Furthermore, the geology of each natural gas well requires a unique set of chemicals to perform the fracking effectively. Therefore, if we are to evaluate whether the chemicals used in a fracking operation is safe or hazardous then we must look at each situation independently.

There is one report which gives a summary of the most often used additives in fracking. The report is titled "Modern Shale Gas Development in the United States: a Primer", produced by the Ground Water Protection Council (GWPC), 2009.

According to the report, some of the most commonly used chemical additives in fracking include: Hydrochloric Acid, Muriatic Acid, Citric Acid, Biocide Glutaraldehyde, Ammonium Persulfate, N-dimethyl formamide, Borate Salts, Polyacrylamide, Hydroxyethyl Cellulose, Potassium Chloride, Ammonium Bisulfite, Silica or Quartz Sand, Ethylene Glycol, and Isopropanol. Some of these chemicals are safe for humans, some are toxic.

In total, the slurry is a 98.0%–99.5% mixture of water, dissolving chemical (acid) and sand. The other additives comprise the remaining 2%.

The Dissolving Acid

The dissolving acid is the additive which is used in the greatest amount. This chemical is usually an acid such as Hydrochloric Acid. The dissolving chemical is typically 15% of the mixture (by volume). This means that the liquid is 85% water and 15% dissolving chemical.

Other Additives

Other additives are then mixed in. Approximately 98% of the final solution (by volume) is water and acid, which means all the other additives combined comprise less than 2%.

The mixture usually contains 3 to 12 additives. These additives perform a variety of functions. Most additives are used to prevent pipe corrosion and reduce friction. Some additives are used to keep the cracks in the rock open (allowing more gas to escape). Other additives assist the natural gas in being pumped to the surface.

Volume of Chemicals Used

The total amount of water required can be as much as 4 million gallons per well. This means the total amount of acids used in a fracking operation ranges from 375,000 gallons to 600,000 gallons per well. The total range of other chemical additives used in a fracking operation ranges from 10,000 gallons to 80,000 gallons per well.

Amount of Chemicals in Waste Water

The waste water from fracking may or may not contain the same amount of additives as the amount sent in. Some of these chemicals will remain in the well. For example, acids will dissolve the rocks and the by-products may remain in the well. Propants will remain in the pores of the rocks, keeping the holes open. Other additives will remain lodged in the pipes.

However, the water will be pumped out again and therefore any of these chemicals can be pumped back up. For example, most of the chlorine and some of the unreacted hydrochloric acid will likely be pumped back up. Therefore the waste water may contain any of the chemicals sent into the well, in varying possible quantities. The only way to know for sure the types and amounts of chemicals in the waste water is to analyze samples periodically.

The chemicals may also be diluted. Additional water is produced from the well (see below), which will dilute the chemicals. Considering the amount of chemicals which remain in the pipes and the well, and taking into account the dilution from additional water produced from the well, the concentration of chemicals in the waste water can be diluted up to 50 times as compared with the concentration of chemicals initially sent in.

In general, we can say the following regarding chemicals in waste water from fracking operations:

a. The waste water from fracking operations will contain some chemicals which are unsuitable for humans or wildlife.

b. The waste water should be tested regularly to determine the types and quantities of chemicals.

c. The waste water from fracking operations should not be deposited into local water supplies or be allowed to contaminate water tables.

Amount of Waste Water from Fracking

The amount of waste water is between 3 million and 4 million gallons. Most of that is from water sent in for the fracking operations. However some additional water is pumped out from the well which came from the well itself. Some of this water comes from underground water tables which have been drilled into. Other water comes from the chemical reaction of acid and rock (creating water and carbon dioxide).

The additional water from underground is an additional 30%–70% of the volume of water put in. A significant negative aspect of this additional water pumped out is that this removed water is no longer part of the underground regional water supply. Like a pipe that keeps leaking, the underground water can continue to leak and be pumped out for a period of months. On the other hand, one positive aspect of this additional water is an extra dilution of the chemical additives in the waste water.

Disposal of Waste Water

As stated above, the waste water from a fracking operation is not useable for the majority of other water uses. The water contains rock, soils, sand, and chemical additives in such large quantities as to be useless in other areas.

For most of the past 60 years, drilling companies have disposed of fracking waste water in one of two ways:

1. Dumping the waste water back into the lake or river
2. Pumping the water into a deep well

Since the creation of the EPA and the environmental movement in general, fewer drilling companies dump the waste water into the local river. However, some drilling companies still do this, so the citizen should be vigilant in watching the operations.

The most common method of waste water disposal today is to pump the waste water into a deep well. The most obvious thought would be to pump the waste water back into the well where the fracking took place. However this requires storing the water until the natural gas is collected, and few companies want to store 4 million gallons of water.

The more common option is to send the water off to an empty well a few miles down the road. This is where some logistics must be considered. Ideally, this well should be close to the fracking well so that the water does not have to be transported far. Some companies are building pipes which connect current fracking operations to other wells. In this way, the waste water is sent directly to the storage well, and this water is sent as soon as it is pumped out from the fracking well.

Each storage well is different, depending on geography and geology, yet the goal of any storage well is the same: the waste water must not contaminate any water used for drinking or agriculture. An ideal storage well will be several thousand feet deep. The walls of the well should be non-porous rock or non-porous industrial material which will contain the water. The storage well should also have special barriers and linings installed so that the waste water does not contaminate the water table and underground water supply.

Injecting waste water into specially designed wells has been common practice for decades. The process is regulated by federal and state agencies. However, there may not be a suitable well for storage of waste water nearby. Furthermore, the amount of water used without additional benefit is enormous. Therefore for both those reasons it is important to develop other options for using fracking waste water.

Uses of Waste Water

Introduction

Up to this point we have 3 million gallons of waste water from each fracking operation in various storage wells. That is a lot of water taken out of the water system. In addition, sometimes there are no suitable storage wells nearby. Therefore in recent years engineers have been evaluating ways to reuse waste water or recycle it.

Methods to Store and Reuse Fracking Waste Water

There are several ways to reuse fracking waste water:

1. Store in an underground well for future use
    a. Re-use in subsequent fracking in same well
    b. Re-use in a second fracking well
2. Store in reservoir on site, using natural filtration
3. Store and use as emergency fire extinguisher
4. Filter for Other Uses
    a. Filter for non-potable, non-agricultural watering of vegetation
    b. Filter for industrial uses
    c. Filter for fire hydrant piping systems
    d. Filter for sending to lakes and rivers
    e. Filter for drinking water

Which method is suitable for use for fracking waste water depends on several factors:

1. Chemical additives in the waste water
2. Composition of dissolved rock and soils in the waste water
3. Degree of purity required for each type of water use
4. Proximity to other places of each type of water use
5. Volume of waste water produced
6. Overall Cost

Use of Wastewater in Subsequent Fracking Operations

The most effective use of waste water is for subsequent fracking operations. The waste water contains much of the sand needed and some of the chemical additives to be used. Minimal filtration is all that is required to remove some of the dissolved rocks. The water is then ready for more chemicals and sand to be added to the solution, which can then be used in another fracking operation.

Natural Filtration On-Site

The waste water can also be stored on site, for any number of uses. A simple filtration system can be implemented using reeds. The water may later become suitable as a natural habitat.

Storing and Using Unfiltered Wastewater

In addition, this water can be used as a supply of water for extinguishing fires. Forest fires and fires in rural areas could be extinguished using this supply of water. Fires from natural gas collected (caused by leaks in containers or pipes) can also be put out with this water.

In general, using fracking waste water for putting out forest fires and prairie fires is an excellent use of this water. The purity of water is not an issue, the goal is to use many gallons of water very quickly. In addition, the sand and soils and the water can be an asset, as these particles can help smother the fire.

Regarding the toxic chemicals in the water, purification may or may not be needed depending on where the fire takes place. Putting out the fire may be more important than the amount of toxic chemicals in the water being used.

Furthermore, the site may require cleaning up anyway (in order to clear out the damaged property) in which case the toxic chemicals from the fracking would be taken away as well.

Filtration Systems for Wastewater

Any other uses of waste water will require a series of filtrations. The amount of filtration required will depend on the final use of the water. Some drilling companies have their own purification systems. Some companies send the water to the local municipal treatment plants.

Best Uses of Fracking Wastewater

The waste water is best used for needs which do not affect health of humans or wildlife. This includes watering grass and cooling industrial equipment, as well as a variety of industrial processes and fire prevention systems.

Waste water should never be put into lakes or used as drinking water, unless that water has been significantly filtered. Some advanced filtration and distillation systems are being developed at this time, but these are recent developments and have not been perfected.

Aquifers and Underground Water Supplies

Overview

The natural gas wells are often located far below water tables, aquifers, water wells, and underground water streams. This presents several problems. The main problems include: the underground fresh water is wasted, drainage causes damage to the water tables, and the underground water supplies may be contaminated with toxic chemicals.

1. Fresh Water is Wasted

As discussed earlier, more water is pumped out of the well than is pumped in. Much of this is essentially from drilling into regions of underground water. A leak is created, and this water leaks for months into the well. This water is wasted: it is not used for any purpose and it becomes contaminated to the degree that it cannot be used in the future without significant purification.

2. Damage to Water Tables

In addition, water ecosystems are very fragile. Draining water from an underground system can cause long-term damage to the entire geology of the region. Generally, when water is completely drained from an underground reservoir the ecosystem above it does not recover. Therefore, if fracking operations remove too much water from the region, the ecosystem may never recover, regardless how much water is eventually put back in.

3. <u>Contamination with Toxic Chemicals</u>

The water tables and aquifers may also be contaminated by the chemicals used in the fracking process. The toxic chemicals and acids used in the fracking process will cross the water tables when going down and when coming back up, making contamination very easy.

The sheer quantity of chemicals used, and the fact that these chemicals are pumped through the underground system both ways, requires a serious look at preventing any pollution to these underground water systems.

The drilling companies protect the underground water systems primarily by using a combination of casing and cement. The inner pipe which sends in blasting solution (and later pumps out waste water and gas) is contained in a second pipe. Therefore, any extraneous solution not in the main pipe will be contained within the outer pipe, and never reach the underground systems.

In addition part of the well is often completely sealed off at a certain level, such that water and chemicals can only move through the piping system. No matter how much extraneous solution (any solution not contained in the pipe) is traveling around in the well, it will never travel up through the sealed barrier.

Note that the depth of the seal is important. If this area is sealed below the water table, then the toxic chemicals are less likely to reach any underground water system.

4. <u>Radioactive Elements in Waste Water</u>

A smaller concern is the issue of radioactive elements. In many rocks there are trace amounts of naturally occurring radioactive elements. The natural gas industry refers to these as "NORM" (naturally occurring radioactive material). The types and amounts of these elements depend entirely on the geology of the particular well.

These elements are generally not a problem because most of these radioactive isotopes exist in small quantities. However, the waste water should be tested throughout the fracking operation for the amount of radioactive decay. If significant amounts of radioactive decay are found in the waste water, then the drilling company must follow the appropriate federal regulations.

Tips and Best Practices for Fracking

There are many practical tips regarding fracking, most of which have discussed above. Here are the essential best practices for fracking:

1. Use chemicals which dissolve rocks yet are not toxic.

2. Never do fracking near aquifers or major water tables.

3. Never add fracking waste water to local streams or lakes.

4. Use filtered waste water in subsequent fracking operations.

5. Limit the use of fresh water for fracking: most of the water for fracking operations should come from previous fracking operations.

6. Purify waste water for use in other operations.

7. Protect the underground water systems in such a way that a) there is no wasted water leaking into the well, and b) the aquifers will not be contaminated from chemical additives.

8. Sample and test waste water regularly for quantities of all toxic chemicals and radioactive decay.

9. Provide the community with clear and honest assessments of health and safety data throughout the fracking process.

Summary of Fracking

1. Fracking is a generic term for a variety of methods for obtaining natural gas from difficult to reach locations.

2. Fracking can be defined as any method which breaks apart rocks deep below the earth's surface so that small pockets of natural gas can be allowed to escape.

3. There are four main methods of fracking. In practice, all four items are used at the same time.
   a. Water: Using water to erode the rocks.
   b. Water Pressure: Using intense water pressure to blast apart rocks.
   c. Chemicals: Using chemicals to dissolve the rocks.
   d. Sanding: Using a type of sand blasting to break apart the rocks.

4. The process of fracking involves sending a liquid slurry under high pressures to dissolve the rocks.

5. After the rocks have been dissolved and broken apart, the natural gas is allowed flow out and be collected.

6. The liquid is a mixture of water, sand, dissolving chemical (usually an acid), and various additional chemicals.

7. The total amount of water required for a fracking operation is between two million and four million gallons. This water is taken from local lakes, underground reservoirs, and municipal water supplies.

8. Due to the significant water requirements for fracking operations, water management authorities must be very judicious when allocating water usage for fracking.

9. The liquid is 85% water and 15% acid. The acid reacts with the rock, creating a hole. By-products include water, carbon dioxide, chlorine, and various compounds from the rock.

10. The remaining chemicals comprise less than 2% of the volume of the mixture. The specific chemicals used depend on the specific geology of the well. Some chemicals are toxic, some are absolutely safe.

11. The total amount of acid used in fracking operations ranges from 375,000 gallons to 600,000 gallons per well.

12. The total amount of other chemical additives used in a fracking operation ranges from 10,000 gallons to 80,000 gallons per well.

13. Waste water from fracking contains sand, rocks, soil, and various chemicals. This water cannot be used for other purposes unless purified.

14. The amount of acid in waste water will be significantly less than the amount sent in because the acid was used to chemically react with the rocks. However elements such as chlorine will be in the same amount.

15. The amount of additional chemicals in the waste water will be less than the amount sent in because many chemicals will remain in the pipes and in the soil.

16. More water will be pumped out than pumped in, up to 70% more. This is because drilling will hit underground water systems. The main advantage is that chemicals in waste water will be greatly diluted. The main disadvantage is that large amounts of fresh water are taken from the region's water supply.

17. Most fracking waste water is deposited into specially designed wells.

18. Fracking waste water can be minimally purified and reused in subsequent fracking operations.

19. Fracking waste water can also be used for fire prevention.

20. Any additional use of waste water from fracking will require significant amounts of purification. The amount of purification required will depend on the final use of the water.

8.5
Refining Natural Gas

Introduction

Overview

Natural gas must be refined in order to be used. Raw natural gas may contain many other molecules besides methane. Therefore, we must separate all the other molecules from the methane.

The amount refining that needs to be done depends on what other molecules are with the methane. Each reservoir, each mine, and each landfill, will have a unique combination of molecules.

However, experience has shown that we can group similar compounds together for the purposes of refining. When these other molecules are looked at in this way, then any source of raw natural gas will contain just a few groupings of compounds.

For the purposes of refining natural gas we can put the possible chemicals of the raw mixture into the following categories:

Grouping Name	Phase
1. Natural Gas	Gas
2. Light Hydrocarbons	Liquids
3. Napthas (also called Condensate, and NGLs)	Liquids
4. Heavy liquid hydrocarbons	Liquids
5. Water, as liquid	Liquids
6. Water, as gas	Gas
7. Sulfur compounds (mostly H_2S)	Gas
8. Minerals, dirt	Solids

General Approach

When refining hydrocarbon fuels we use common physical properties for separation, particularly phase, density, and boiling point. If there are several molecules with similar physical properties, then the molecules which exist in greatest quantities are separated first.

The general order of operations for separating all other molecules from the methane is as follows:

1. Remove water
2. Remove solids, large size
3. Remove hydrocarbon liquids
4. Remove gas phase molecules:
    a. Remove hydrocarbon gasses
    b. Remove hydrogen sulfide
    c. Remove water (gas phase)
5. Remove other molecules

Removing Water and Solids

Removing Water – as Liquid

The first molecule to remove is water in the liquid phase. Water is removed by the differences in density. Water is more dense than most hydrocarbons. When the mixture is allowed to settle in a container, the hydrocarbons will rise to the top, and the water will settle to the bottom. The hydrocarbons can then be collected for further refining. Note that there will still be water in the gas phase which will be removed later.

Removing Solids – of Larger Size

Many solid particles might also be in the mixture. These solids can be separated using the same technique above. When the mixture is allowed to sit, the solid particles will settle to the bottom. Note that there will still be small sized particles in the mixture. These will be removed later.

Removing Hydrocarbon Liquids
including most Oils and Condensate

Many sources of methane will have liquid hydrocarbons. In general, these liquid hydrocarbons will have between 5 carbons and 40 carbons in the chain.

The lighter of these hydrocarbon liquids might be referred to as Napthas, Natural Gas Liquids, or Condensate. These terms generally apply to liquid hydrocarbons with 5-7 Carbons in the molecule.

The heavier of these hydrocarbon liquids might be referred to as Gasoline, Fuel Oil, or simply "Oil." These terms generally apply to liquid hydrocarbons with 7-40 Carbons in the molecule.

The most common method of removing liquid hydrocarbons from the raw natural gas is a "conventional separator." In this device, the molecules separate by gravity. The molecules in the oil and in the condensate are more dense than methane, so they settle to the bottom. Any hydrocarbons in the gas phase (including methane molecules) rise to the top.

The methane (and other gas molecules) will be collected and taken away for further refining. The heavier hydrocarbons are also taken away for further refining, in a different set of processes.

Removing Hydrocarbon Gasses

Overview

After all of the hydrocarbon liquids are removed there are several hydrocarbon gasses which remain. These gases include methane, ethane, propane, and butane. Each of these light hydrocarbons is valuable, therefore removing these chemicals is not only cost-effective, but profitable.

There are two primary methods used to separate the remaining hydrocarbon gasses from methane: use an absorbing oil or use a lower temperature.

1. Absorbing Oil

In this method a special absorbing oil is sent in. The oil is a particular mixture of liquid hydrocarbons which was blended just for this purpose.

Each of the remaining liquid hydrocarbons is attracted to this absorbing oil, and the liquid hydrocarbons grab onto that absorbing oil. This absorbing oil, and the hydrocarbons which grabbed onto it, can then be taken away.

2. Lowering Temperature

When we lower the temperature of the mixture, more of the gas phase molecules will become liquid. This means that more of the ethane, propane, and butane will collect as liquid. We can then separate these liquids from the methane gas. Note that we can cool the temperature to any amount as long as we don't lower the temperature low enough for the methane to become liquid.

Removing Sulfur (Removing Hydrogen Sulfide)

When sulfur exists in raw natural gas the sulfur usually exists as hydrogen sulfide in the gas phase. The sulfur is removed by the addition of an amine solution. The amine solution is usually monoethanolamine (MEA) or diethanolamine (DEA). This amine solution is added to the raw natural gas, then the sulfur is attracted to the amine solution. When the amine solution is taken away the sulfur is taken with it.

Each of these amine solutions can be recycled. The sulfur is removed from the amine solution, then the amine solution can be sent into the next batch of raw natural gas. In addition, the sulfur can then be sold to various industries. Note that 90% of sulfur production in the U.S. is created by this method, 15% of which is obtained specifically from raw natural gas.

Removing Water in the Gas Phase

Overview

Although the liquid water has been removed in the first step, there is still a significant volume of water in the vapor phase. Depending on the temperature of any segment of pipeline the water can become liquid or solid, and therefore impede the flow of natural gas in the pipelines. Therefore the remaining water must be removed before the natural gas is distributed. There are two primary methods for removing water in the gas phase: glycol absorption, and adsorbing particles.

Note that when using these methods some water in the gas phase usually remains. The natural gas usually needs to be sent through water removal treatment several times. This is done at compression stations along transmission pipelines.

1. Glycol Absorption

The most common method of removing water vapor from methane is to use glycol absorption. In this method a glycol solution is sent into the raw natural gas. Glycol grabs onto the water. These glycol–water molecules sink to the bottom and then can be removed. The glycol solution is usually diethylene glycol (DEG) or triethylene glycol (TEG).

Either of these solutions can be reused. In order to reuse the glycol solution, the water must be separated from the glycol. The water can be removed from the glycol solution using the difference in boiling points.

2. Adsorbing Particles

Water vapor can also be separated from methane by using particles which adsorb the water. (See adsorb vs. absorb in the Unit on Coal.) In this method, particles are put into the stream of natural gas. These particles are usually activated alumina or silica gel. Water molecules adsorb onto the particles. The gas stream is then purified of water.

This method tends to capture more water molecules than the glycol solution method. However, the particles will reach a limit of adsorption, and therefore must be replaced.

The adsorbing particles can be removed and heated in order to empty all the water molecules. These particles can then be reused again. However, in the meantime there must be a second set of adsorbing particles in the natural gas stream so that water molecules continue to be adsorbed.

Removing Other Molecules (trace, inert)

If all of the other molecules above have been removed then the natural gas should be pure methane. However, there may be several other molecules which may exist in the natural gas. Some of these other molecules include helium, carbon dioxide, and nitrogen. Sometimes inert gasses exist such as well.

Solids may also exist. Just as fly ash exists in the flue after coal combustion, tiny particles can exist in the pipelines carrying natural gas. These tiny solids might include coal dust (if the methane was taken from a coal mine), dirt, and various minerals.

These small solid molecules are usually safe for people and the environment. However, these molecules can have practical importance. First, these other molecules can impede the flow of the methane. Second, because these molecules do not burn, these molecules make the final burning of natural gas much less efficient.

Most of these chemicals are removed at compression stations along the pipeline. These molecules are difficult to separate, therefore the techniques involved are more subtle. (Further details of the techniques are beyond the scope of this book.)

Summary of Natural Gas Refining

1. Natural gas must be refined in order to be used. The amount refining that needs to be done depends on what other molecules are extracted with the methane.

2. When refining hydrocarbon fuels, the most common physical properties used are phase, density, and boiling point.

3. The general order of separation is as follows:
    a. Remove water
    b. Remove solids, large size
    c. Remove hydrocarbon liquids
    d. Remove gas phase molecules
    e. Remove other molecules

8.6
Transporting and Storing Natural Gas

Introduction

Overview

If we want to use natural gas for creating electricity then we must transport it. Whereas other fuels can be delivered by truck, natural gas is most effectively delivered by pipelines.

There is already an extensive pipeline infrastructure in place throughout the nation. In addition, pipelines continue to be installed. However, there are several technical details involved in order to make this system work.

Three Pipeline Systems

The infrastructure of pipelines for natural gas is divided into three systems: Gathering, Transmission, and Distribution. Each system is a distinct set of pipelines, and each has its own requirements.

1. The Gathering System carries raw natural gas from the original location to a refinery.

2. The Transmission System carries purified natural gas in large quantities to regional locations (such as to major cities). Note that most of the power plants which use natural gas are connected to the transmission pipeline system.

3. The Distribution System carries purified natural gas in smaller quantities to individual customers.

Each of these systems differs in practical ways. The important specifics of each system will be discussed in subsequent sections of this chapter.

Compressors

In order to send gas through the pipes, the gas must be pushed periodically. This push is done by compressors. The gas must be compressed at least every 75 miles. Further details on compressors will be discussed in a later section of this chapter.

Liquefied Natural Gas

There is an alternate method to the pipeline system. This method is turning Natural Gas into Liquefied Natural Gas. In brief, this means we cool the gas into a liquid. Then the fuel is transported in metal containers aboard ships and trucks. This method will be discussed in more detail in a subsequent section.

Gathering System

Introduction

The Gathering System carries raw natural gas from the original location to a nearby refinery. Note that it is important to build refineries near the sources so that the hazardous mixture of raw fuel does not have to travel very far. The pipes must have special coatings which resist chemical attacks from the hydrogen sulfide. The pipes are generally small in diameter, and the gas is usually low pressure.

It is important to remember that the gathering system carries raw fuel which may be a mixture of many chemicals. Therefore the mixture sent through the gathering system can be very dangerous. The most important hazards include the following: hydrogen sulfide, flammable hydrocarbons, carcinogenic molecules, and toxic molecules.

Hydrogen Sulfide (H_2S)

Hydrogen Sulfide is very toxic, it is in the gas phase and exists in high concentrations. Any leak will be very harmful, possibly deadly, to anyone nearby.

Flammable Hydrocarbons

Raw natural gas and crude oil may contain dozens, if not hundreds, of hydrocarbons. Many of these will be later refined into fuel. Remember that some of these fuels include: methane, propane, gasoline, and diesel gas. Therefore, this mixture can and will burn given the right conditions.

Carcinogenic and Toxic Carbon-Chain Molecules

Within a mixture such as crude oil there are often many chemicals which are toxic or carcinogenic. These chemicals may harm humans or the environment and therefore should be handled carefully.

Transmission System

Introduction

The Transmission System for natural gas exists to carry purified natural gas from the refineries to various distribution centers. After the natural gas has been purified (into pure methane) then the natural gas is ready to be sent across the country for use. This task is done via the Transmission System.

Hazards of Transmission System

The fire hazard along transmission pipes is minimal. The transmission pipes are located outdoors, often along less populated routes. Therefore, if there is a leak, the gas will spread up and out to the air. There is less fuel in one spot and therefore less fuel to possibly burn.

However, there will always be a potential for fire from transmission pipes, therefore these pipes must be inspected regularly for leaks. Leaks are usually identified visually. When methane leaks from the pipe the methane will rise up through the ground and therefore the ground on the surface will become discolored. Leaks can also be identified using a robotic inspection tool. Note that the natural gas in the transmission pipes does not usually have an odor. Therefore leaks cannot be identified by smell.

Pipelines for Transmission System

There are two categories of transmission pipes: main lines and laterals. A main line is similar to a freeway: a large quantity of gas is sent through the pipe, at high speeds, with very few exit points. The main lines are between 16 inches and 48 inches in diameter. The gas in the transmission pipelines is pushed at very high pressures, usually between 200 psi and 1,500 psi.

Lateral pipelines lead to or from the main lines. Lateral pipelines are used as entry points and exit points. Laterals are most commonly used to send natural gas from the main transmission pipeline to the local distribution centers. Lateral pipelines are also used to send gas from the main line to each natural gas power plant. Lateral pipelines are between 6 inches and 16 inches in diameter. The pressure in the lateral is lower than in the main line.

In order to move the gas through the pipes the gas must be compressed at frequent intervals. This is done at compressor stations. After the gas is compressed, the gas will travel approximately 15-30 miles per hour. The gas will gradually slow down until it reaches the next compressor.

Federal law (from the Office of Pipeline Safety) mandates that transmission pipelines be buried to a sufficient depth. The top of the transmission pipeline must be 30 inches below the ground. The pipeline must be buried deeper than the minimum for factors such as under streams, under roads, and more populated areas. The total depth of the trench for transmission lines is usually 5 or 6 feet.

Distribution System

Introduction

The Distribution System carries purified natural gas, in small quantities, from distribution stations to individual customers. Note that most natural gas power plants get their fuel from the Transmission pipelines not from the Distribution pipelines. Therefore, because the focus of this book is electricity we will be very brief regarding the Distribution System of natural gas. On the other hand there are some small power plants which do indeed obtain their supply of natural gas from the Distribution System. Therefore, some discussion of the Distribution System is necessary.

Regional Distribution Center: Citygate, Gate Stations

The regional distribution center is known as the "citygate." This is because the central storage area for the region is usually just outside the city. These centers are also called "gate stations." At the citygate, several things happen:

    1. The pipeline ownership changes
    2. Natural gas is stored for later use
    3. Natural gas is purified to a greater extent
    4. Natural gas is reduced in pressure
    5. An odor is added to the natural gas

Ownership: Local Distribution Companies (LDCs)

Delivery of natural gas within a city is considered a natural monopoly. This is because we don't want too many pipelines congesting the same area. The owners of the distribution pipeline system are called Local Distribution Companies, often abbreviated to LDCs.

Distribution companies for natural gas usually operate within a distinct geographic area. Distribution pipelines are regulated by the Federal government, mostly by the Office of Pipeline Safety (within the Department of Transportation) and by the Federal Energy Regulatory Commission. Distribution pipelines are also usually regulated by each state.

Storing Natural Gas

Natural gas can be stored until needed. There are several requirements for storing natural gas: 1) large enough space to store the needed gas, 2) no leaks in the container, 3) no sources of flame, sparks, or heat nearby, and 4) storage is away from most of the population.

Natural gas has been most commonly stored underground. In many cases, depleted wells and mines are used as storage facilities. There are several advantages to this method:

- The space is large enough to store large volumes of gas
- The space is far from flames or sparks
- The location is usually far enough from populated regions
- The location is not likely to be used for other projects

Additional Purification and Addition of Mercaptans

Although the natural gas sent to the distribution company is almost pure methane, there may be other elements. Therefore the local distribution company often purifies the natural gas before sending it to customers. Also, methane has no color or odor which makes it impossible to detect. An odor is added so that customers can smell leaks.

The most common odors added are sulfur compounds known as thiols. Thiols are commonly referred to as Mercaptans. A thiol is similar to an alcohol, but with the oxygen atom replaced with a sulfur atom. For example, ethanol (an alcohol) is C_2H_5OH. The related thiol is ethanethiol, C_2H_5SH.

There are numerous chemicals which are classified as a thiol (or as a mercaptan.) Different thiols have different odors. Each distribution company selects its own thiol (or blend of thiols) to add to its natural gas.

<u>Pipelines for Distribution System</u>

The network of pipelines for the Distribution System is quite extensive. These pipes carry small volumes of gas over short distances to numerous individual customers. There are over 1 million miles of distribution pipes throughout the United States.

The pipes for the distribution system are made of steel or plastic. Steel is used where the pressure is greater than 100 psi. Plastic is used where the pressure is less than 100 psi. Most of the pipes within the distribution system operate at low enough pressure to be made of plastic. The pressure ranges from 200 psi (when leaving the citygate), down to merely a few psi in most distribution pipes, and as little as 1/4 psi in the homes.

The natural gas must be compressed periodically, however the compressor stations are much smaller than the compressor stations used for transmission pipeline systems.

Compressors

<u>Introduction</u>

In order to send gas through the pipes the gas must be pushed periodically. This push is done by compressors. As the name implies, compressors apply pressure on the gas, resulting in a strong push. In order to keep the gas flowing down the pipe, compressors must be used every 50–75 miles.

Compressors are often very large, usually housed in a small building. Compressors are also very loud, with the sound comparable to a jet engine.

The act of compressing is done by a turbine, a motor, or an engine. Each device performs the job differently, though the net result is the same.

Note that in order to do the act of compressing some energy must be used. This energy is often taken from the natural gas itself.

Separators at Compressor Stations

Some of the compressor stations have a separator as well as the compressor. The primary function of the separator is to remove any remaining water from the gas. (This may require several iterations at several compressors down the line).

Additional separators are installed at some compressor stations to remove other molecules. These other molecules are usually gasses, which are difficult to remove completely at the refinery.

Corrosion Resistance for Natural Gas Pipelines

Introduction

Transmission pipelines are usually made of steel. Steel is necessary when the pressure of the gas is over 100 psi. However steel can corrode in many soils, which will result in leaks. Therefore it is important for gas pipelines to be resistant to corrosion.

There are two general methods for corrosion resistance for steel pipes. Usually both methods are used because one method alone is insufficient. These methods are: 1) coatings and 2) cathodic protection.

Coatings

All transmission pipelines are required to have coating which resists corrosion. A coating works simply by not reacting with water or ions in the soil. There are many materials which can provide corrosion resistance. Some of these include: aluminum, zinc, plastics, and rubber. The most common coating for transmission steel pipes is a fusion bond epoxy (FBE).

However, no coating method is will able to cover the pipe 100%. There will always be a few holes in the coating. Corrosion will then occur at these holes. Therefore another protection method is necessary, such as cathodic protection.

Cathodic Protection

In addition to special coatings, corrosion resistance can be done through cathodic protection. Cathodic protection uses an external power source, usually a small battery, to counteract the corrosion process.

Additionally, the cathode protection device often has an extension above ground so that workers can check on the status of corrosion and to replace the battery when it is depleted. In some situations the battery for cathodic protection can be recharged by solar power.

Natural Gas in Power Plants

Introduction

The infrastructure of extraction, purification, and transportation of natural gas has long been in place for heating and cooking in homes. This same infrastructure allows us to use natural gas for the production of electricity.

Most of the large gas-fired power plants get their fuel from the transmission pipeline system. As stated earlier, the transmission pipeline system carries large volumes of natural gas. Therefore if a large power plant is to be built the power plant should connect to these transmission lines.

The turbines used for natural gas power plants can be gas, steam, or combined cycle. Large natural gas power plants typically use a combined cycle system, where the gas turbine is used first followed by the steam turbine.

Hazards of Using Transmission Pipelines

It is important to remember that the gas in the transmission pipeline does not have any odor, therefore if there is a leak then the gas will not be detected by smell. Also remember that the final pipes will lead somewhere indoors as the gas reaches the turbines. Therefore if there is an undetected leak then a large amount of gas could build up in the room. Heat within the building or a small spark from a machine can ignite the natural gas. This could result in a fire or explosion. Therefore, power plants which use natural gas must have methane detectors, and operators must inspect all pipes for leaks frequently.

Liquefied Natural Gas (LNG)

Introduction

Liquefied Natural Gas (LNG) is an alternative to shipping gas through pipelines. Liquefied Natural Gas is also an alternative method to storing gas in underground caverns.

Liquefied Natural Gas is essentially natural gas which has been cooled into liquid, and kept cool throughout shipping. The natural gas must be cooled to –260 °F in order to become liquid.

The advantage of Liquefied Natural Gas is that it takes less space than normal natural gas. Liquefied Natural Gas takes only 1/600 the volume of the same mass as it existed in the gas phase. For example, 600 cubic feet of natural gas can be cooled and condensed into one cubic foot of liquefied natural gas. Because the natural gas takes up less space, it can be shipped more efficiently. This process is most effective for natural gas being shipped across oceans.

When choosing between traditional pipelines or liquefied natural gas as the method for transportation the factors are cost and simplicity. The liquefied natural gas will be used where it is cheaper or easier for the particular circumstances.

Safety of LNG

An additional concern for storing Liquefied Natural Gas (which is not a concern for regular natural gas) is to make sure that the methane does not become gas while in the container. If the liquid methane warms up enough to become gas then the container will burst. Remember that the temperature for methane to become gas is very low (–263 °F), so it will not take much heat to turn the liquid into gas.

In order to transport or store Liquid Natural Gas safely the following requirements must be met:

1. The container for LNG must have sufficient insulation.
2. The container must only be partially filled.
3. The container must use multiple layers of containment.
4. There must be no leaks in the container.
5. There must be no sources of flame or sparks nearby.
6. The container should be stored away from heat.
7. The container must be stored away from populated areas.

The U.S. Coast Guard is the primary agency which supervises the transportation of Liquefied Natural Gas. This is because most LNG containers are shipped on large tankers across the ocean.

Another regulating authority is the Department of Transportation. Their requirements for transporting Liquefied Natural Gas are similar to the requirements for transporting oil and gasoline.

Summary of Transporting Natural Gas

1. Natural gas is most effectively delivered by an extensive network of pipelines.

2. The infrastructure of pipelines for natural gas is divided into three systems: Gathering Pipeline System, Transmission Pipeline System, and Distribution Pipeline System.

3. The Gathering System carries raw natural gas from the original location to a refinery.

4. The pipes in the gathering system are generally small in diameter and the gas is carried at low pressure.

5. The pipes in the gathering system must have special coatings which resist chemical attacks from the Hydrogen Sulfide.

6. Refineries should be built near the sources of natural gas so that the hazardous mixture of raw fuel does not have to travel very far.

7. Because the gathering system carries raw fuel, the mixture can be very dangerous. The hazards in the gathering system include toxic hydrogen sulfide, flammable hydrocarbons, and carcinogenic molecules.

8. The Transmission System carries purified natural gas in large quantities to regional locations.

9. The transmission pipes are generally wide, up to 48 inches in diameter. The transmission pipelines carry natural gas at very high pressures, up to 1,500 psi.

10. Although methane is very flammable, the fire hazard along transmission pipes is minimal. If there is a leak in the transmission pipe the gas will spread into the air. However, there will always be a potential for fire, therefore transmission pipes must be buried to a sufficient depth, must be resistant to corrosion, and must be inspected regularly.

11. Leaks in the transmission pipelines are usually identified visually. When methane leaks the ground becomes discolored. Natural gas in the a transmission pipe does not usually have an odor. Therefore leaks cannot be identified by smell.

12. Transmission pipelines are usually made of steel. However, steel can corrode which will then produce leaks. There are two general methods for corrosion resistance for steel pipes: coatings and cathodic protection.

13. The Distribution System carries purified natural gas in small quantities to individual customers.

14. The regional distribution center is known as the citygate or the gate station. At the citygate, several things happen:
    a. Natural gas is stored for later use
    b. Natural gas is purified to a greater extent
    c. Natural gas is reduced in pressure
    d. An odor is added to the natural gas
    e. The pipeline ownership changes

15. The owners of the distribution pipeline systems are called Local Distribution Companies, often abbreviated to "LDCs".

16. The pipes for the distribution system are made of steel or plastic. Steel is used where the pressure is greater than 100 psi. Plastic is used where the pressure is less than 100 psi.

17. The diameter of distribution pipes usually range from 24 inches down to 1 inch. The pressure ranges from 200 psi (leaving the citygate) down to as low as 1/4 psi in the homes.

18. Natural gas can be stored until needed. There are three requirements for storing natural gas:
    a. Large enough space to store the needed gas
    b. No leaks in the container
    c. No sources of flame, sparks, or heat nearby
    d. Storage is away from most of the population

19. Natural gas has been most commonly stored underground, usually in depleted wells and mines. There are several advantages to this method:
- The space is large enough to store large volumes of gas
- The space is far from flames or sparks
- The location is usually far enough from populated regions
- The location is not likely to be used for other projects

20. Methane has no color or odor. An odor is added at the distribution stage so that customers can smell leaks.

21. To send gas through the pipes the gas must be pushed periodically. This push is done by compressors.

22. Compressors are used in both transmission systems and distribution systems. The size of the compressors depends on the volume of gas which needs to be compressed, and the desired pressure.

23. Some of the compressor stations have a separator. The primary function of the separator is to remove any remaining water from the gas.

24. An alternate method to transporting natural gas in the pipeline system is to cool the methane into Liquefied Natural Gas (LNG). The advantage of Liquefied Natural Gas is that it takes 1/600 the volume of normal natural gas. The LNG can be put into special containers and transported on ships.

25. In order to transport and store Liquid Natural Gas safely, the following requirements must be met:
   a. The container for LNG must have sufficient insulation.
   b. The container must only be partially filled.
   c. The container must use multiple layers of containment.
   d. There must be no leaks in the container.
   e. There must be no sources of flame or sparks nearby.
   f. The container should be stored away from heat.
   g. The container must be stored away from populated areas.

26. Most of the large gas-fired power plants get their fuel from the transmission pipeline system.

27. Connecting to the transmission pipeline system is effective for a large power plant because of the large volumes of gas delivered. However, there is no odor in this gas so leaks cannot be detected by smell. Alternate leak detections must be used, along with frequent inspections.

28. The turbines used for natural gas power plants can be gas, steam, or combined cycle. Larger plants should use combined cycle turbines.

29. Natural gas is a very clean fuel. When pure natural gas is burned the by-products are only carbon dioxide and water.

8.7
Other Hydrocarbon Fuels

Introduction

In this chapter we will discuss the final sources of energy commonly used in the generation of electricity. These fuels are diesel fuel, biomass, and cogeneration.

Diesel Fuel for Electricity

Diesel gasoline is an important fuel for electrical power because diesel generators are ideal for emergency back-up systems. Diesel fuel can be stored virtually forever, and therefore the fuel is always ready in an emergency. The fuel is also portable, which allows the generator and fuel to be taken anywhere.

Diesel generators typically use combustion (gas) turbines. When diesel fuel is burned, the result is carbon dioxide and water in the gas phase. These gasses push on the turbine, and the turbine operates the generator.

The most significant advantages of using diesel fuel for electrical generators are the following:
1. Diesel fuel can be stored.
2. Diesel generators are quick to start up.
3. Diesel generators can be portable.
4. Diesel generators are excellent for emergency power.

The most significant disadvantages of using diesel fuel are the following:
1. Diesel fuel is a non-renewable resource.
2. Diesel fuel can be harmful to people and to the environment.
3. Diesel generators are noisy.

Diesel Fuel Science

Introduction

Diesel fuel is a mixture of liquid hydrocarbons. The specific mixture depends on what the particular company decides to use. The most specific we can be about any diesel fuel is as follows: Diesel fuel is a mixture of hydrocarbons, where any hydrocarbon in the mixture has 10 to 20 carbons in its chain.

The primary reason for using diesel fuel is the amount of energy. Diesel fuel has the most energy per volume of all the liquid hydrocarbons. For example, typical diesel fuel has 18% more energy than the same volume of gasoline. This is the reason why diesel fuel is often used for large equipment such as trucks and trains.

Diesel fuel is also easy to store. As a liquid, diesel fuel can be stored for long periods of time. Diesel fuel is also much easier, and safer, to store than liquefied gases (such as liquefied propane or liquefied natural gas).

Carbon Particles in Diesel Fuel

Diesel fuel is often very dirty. This is primarily due to the many particles which exist in the diesel mixture and do not burn.

Most of these particles are higher chain carbon molecules, essentially pieces of tar or coal. Remember that diesel fuel is composed of higher chain molecules, and remember that tar and coal are not that much further up the sequence of hydrocarbon chains. Therefore it is common for small tar or coal like pieces to form within the diesel fuel.

These particles will remain as part of the diesel fuel mixture, including throughout the storage and use of the diesel fuel. Consequently these particles will clog the pipes – which restricts the flow of fuel. These particles will also not burn efficiently – which results in many particles being emitted into the air.

Of course the fuel can be refined to eliminate those particles. The degree of purification depends on the refinery. However, most diesel fuel currently available will contain these coal and tar particles.

Other Pollutants in Diesel Fuel

Diesel fuel also has significant quantities of sulfur, more than regular gasoline. (This is another way diesel is like coal: both coal and diesel tend to have large quantities of sulfur). Diesel fuel may also have significant quantities of carcinogenic organic molecules. In addition, the impurities in diesel fuel may be any of the impurities discussed in the chapters on natural gas.

Purification of Diesel Fuel

Diesel fuel can be purified. Many of the processes described for the purification of natural gas can also be applied to purifying diesel fuel.

Another alternative is to use biodiesel instead of traditional diesel. Biodiesel has far fewer pollutants than traditional diesel, yet the same hydrocarbon chemicals.

Biodiesel

Introduction

Biodiesel fuel is an alternative to traditional diesel fuel. The desired fuel is the same in both diesel mixtures: a mixture of hydrocarbons each of which has 10-20 molecules. However, the method of obtaining the fuel is very different. The net result is a mixture which has more of the desired fuel and less of the harmful chemicals.

Making biodiesel is a complex process, and is beyond the scope of the book. However we can briefly describe the process here. Biodiesel begins with vegetable oils or animal fats. The vegetable oils are combined with methoxide, then heated. Sometimes a catalyst is added. One of the products will be biodiesel fuel. Another product will be a form of soap. Other products may include glycerol and types of alcohols. Each product is separated, collected, and used.

Efficiency of Diesel Production

The efficiency for the production of biodiesel is usually measured in terms of "the amount of biodiesel fuel created per mass of starting material". Two of the most effective sources for starting material are algae and palm oil. Waste vegetable oils are also popular.

Many people encourage the use of Waste Vegetable Oils (WVO). Waste Vegetable Oils may not be the most efficient starting material for the creation of diesel fuel, but these oils are readily available. In this regard, using WVOs are "efficient" because the starting material is easy to obtain.

Comparison to Traditional Diesel

Biodiesel has no sulfur. This is in sharp contrast to traditional diesel which has significant amounts of sulfur. Biodiesel is also non-toxic. This is in contrast to traditional diesel which may have a number of toxic chemicals in the mixture. Biodiesel also burns more efficiently, producing less carbon monoxide than traditional diesel fuel.

Biodiesel is not yet a cost effective alternative to traditional diesel. However, it is becoming more common to mix some biodiesel with traditional diesel. This reduces the overall amount of pollutants in the mixture (as compared to traditional diesel alone). Mixing biodiesel with regular diesel also helps with the job of lubricating, and therefore fewer lubricating additives are needed.

Biodiesel is more commonly sold as a mixture than as pure biodiesel. Mixtures are labeled "B" for Biodiesel, followed by a %. For example, "B20" tells us that the diesel fuel is 20% Biodiesel.

Biomass and Trash

Introduction

Biomass is any dead plant material which we burn in order to create heat. The most common forms of biomass are firewood and agricultural waste. Trash for this process can be many items, but it is best if the items are mostly carbon based (wood, food, some construction materials).

The basic process for getting electricity from biomass or trash is relatively simple. We burn the biomass (or trash) which creates heat. This heat is then used to boil water, which creates steam. The steam operates the turbine, and the rotating turbine operates the electrical generator.

Burning any biomass or trash to generate electricity offers several benefits: we eliminate waste, we can create heat, and we can create electrical power. However, not all trash and biomass are suitable for burning. Some types of trash may become toxic when burned, and some types of biomass may be put to better use as compost.

Types of Biomass for Burning

There are many specific biomass items that can be used for burning. Some examples include: firewood, wood scraps, leaves, twigs, dead branches, fallen trees, cornstalks, sugar canes, and waste from any agricultural process.

Types of Trash for Burning

There are many specific trash items that can be used for burning. The best trash items for this purpose are those made of carbon chains. Some examples include paper, cloth, and food trash. A good rule of thumb is the following: if the trash item came from nature, or if the item is biodegradable, then it may be suitable for burning.

Limitations on Burning Biomass and Trash

If we are to burn biomass or trash as a source of power generation then we must understand two limitations. The first limitation is the act of burning biomass or trash must be done close to customers. The main reason is that the energy we get from burning biomass or trash is relatively low compared to the other traditional sources of energy.

Furthermore, if these wastes were to be transported to a power plant site, this transportation will add to the cost of the electricity.

The second limitation is that not everything can be burned. This is because the items are toxic, explosive, or do not burn effectively.

Cogeneration

Cogeneration is the process of using the same fuel for the generation of electricity and the creation of heat. Cogeneration is a process done by large facilities, where the power is both generated and used at the same site. Cogeneration is done mostly for greater efficiency and for greater independence.

As a simple example, consider the process of burning biomass on a farm. The stalks and leaves can be burned. When this organic material is burned, heat is created. That heat is used to keep the buildings on the farm warm. Furthermore, the burning of biomass creates molecules in the gas phase (mostly CO_2 and H_2O). These gas phase molecules can be used in a combustion turbine in order to create electricity for the farm. The net result is both heat and electricity created by the same fuel.

Specific advantages of cogeneration include the following:

- Cogeneration allows us to heat more efficiently.
- Cogeneration allows users to be more independent.
- Cogeneration eliminates waste effectively.

Cogeneration equipment comes in a variety of forms. However, the one thing all of these generators have in common is the creation of both heat and electricity from the same fuel. The fuel for cogeneration is usually oil, natural gas, trash, or biomass.

Cogeneration can be used in buildings of all sizes. However, due to logistics (such as cost, maintenance, and daily operation) cogeneration is best suited where many people work under one management. Therefore, cogeneration is especially useful for large businesses, farms, schools, and universities.

Summary of Other Hydrocarbon Fuels

1. Diesel fuel is a mixture of hydrocarbons, each with 10-20 carbons in its chain.

2. The primary reason for using diesel fuel is the amount of energy. Diesel has the most energy per volume of all the liquid hydrocarbons.

3. Diesel generators are ideal for emergency back-up systems.

4. Diesel fuel is often very dirty. This is due to carbon particles, sulfur, and toxic organic molecules in the mixture.

5. Diesel fuel can be purified to various degrees. However, due to business reasons most of the diesel fuel available is less purified than is technologically possible.

6. Biodiesel fuel is an alternative to traditional diesel fuel. The desired fuel is the same, but there are fewer harmful chemicals in biodiesel than in ordinary diesel.

7. Biodiesel is more commonly sold as a mixture than as pure biodiesel. Biodiesel blends are noted by "B" and the % biodiesel.

8. Advantages of diesel fuel for electricity include:
    a. Diesel fuel can be stored for long periods of time.
    b. Diesel generators are quick to start up.
    c. Diesel generators can be portable.
    d. Diesel generators are excellent for emergency power.

9. Disadvantages of using diesel fuel include:
    a. Diesel fuel is a non-renewable resource.
    b. The diesel fuel mixture can be harmful to people and to the environment.
    c. Diesel generators are noisy.

10. Biomass is any dead plant material which we burn in order to create heat. The most common forms of biomass are firewood and agricultural waste.

11. The basic process for getting electricity from biomass is relatively simple. We burn the biomass which creates heat. This heat is then used to boil water which creates steam. The steam operates the turbine, and the rotating turbine operates the electrical generator.

12. Trash can be burned in the same process as biomass in order to create electricity. The best trash items for this purpose are those made of carbon chains.

13. There are two limitations on burning biomass or trash for generating electricity:
    a. Burning biomass or trash must be done close to customers.
    b. Not everything can be burned.

14. Burning biomass and trash can be good, however some items are best disposed of in other ways or be used for other purposes.

15. Cogeneration is the process of using the same fuel for two purposes: the generation of electricity and the creation of heat. In general we choose to use cogeneration for greater efficiency and independence.

16. Cogeneration equipment comes in a variety of forms. The one thing all these generators have in common is the creation of both heat and electricity from the same fuel. The fuel for cogeneration is usually oil, natural gas, trash, or biomass.

17. Cogeneration can be used in buildings of all sizes. However, cogeneration is especially useful for large businesses, farms, schools, and universities.

Conclusion

Many Americans hold passionate views about electrical power, yet few Americans understand all the details behind their passion. Electricity should not be mysterious. The science, the technology, and the data of electrical power can be understood by anyone.

Above all else, we must remember that there are no perfect solutions, there are only choices. Any option can be beneficial, yet each option has its own technical issues to work with. It is up to you and to your community to make those educated decisions. I hope that this book will help guide you in your choices.

M.F.

Appendix

1. <u>Terms related to Natural Gas</u>

The following terms are the most common terms which you will find when reading about natural gas.

<u>Associated Gas</u> – Associated gas is natural gas which is found with oil.

<u>Condensate</u> – Condensate is any liquid which exists with the natural gas. Condensate usually includes one or more of several light, liquid hydrocarbons. Note that oil is not usually considered a condensate because oil is made of heavier liquid hydrocarbons.

<u>Condensate Well</u> – A condensate well is a natural gas well which has liquid hydrocarbons other than oil. The well contains primarily natural gas. There is little or no oil. However, the well has other hydrocarbons (other than oil or methane) and these hydrocarbons are usually in the liquid phase.

<u>Dry Natural Gas</u> – Dry natural gas is essentially pure methane. Natural gas is considered "dry" not merely by the removal of water, but by the removal of all the other hydrocarbons. Remember that raw natural gas often contains several hydrocarbons. Pure natural gas (methane) exists as a gas. Many of the other hydrocarbons exist as liquids. Therefore, "dry natural gas" is pure methane, after the natural gas has been refined.

<u>Landfill Gas</u> – Landfill gas is the mixture of all gas phase molecules naturally being emitted from a landfill.

<u>Most Efficient recovery Rate (MER)</u> – The MER is the rate at which the greatest amount of natural gas may be extracted without harming the formation. (The formation is the natural barrier which contains the natural gas).

Natural Gas Liquids (NGLs) – Natural Gas Liquids (NGLs) are liquid hydrocarbons which exist with the natural gas. NGLs are also called Condensate and Napthas. These terms generally apply to liquid hydrocarbons with 5-7 Carbons.

Pipeline Inspection Tool (PIG) – The Pipeline Inspection Tool is a robotic device which is sent into the pipes. This tool is often called a "PIG", for Pipeline Inspection Gadget. The PIG inspects the interior of the pipe. Some of the inspections include: pipe thickness, pipe roundness, corrosion, and leaks.

Raw Natural Gas – Raw natural gas is the natural gas just after it is extracted from the ground. Raw natural gas contains many other molecules besides the desired methane.

Shale – Shale is a type of rock similar to clay which has very little permeability. Pockets of natural gas are often trapped behind the shale.

Wet Natural Gas – Wet natural gas is essentially the same as raw natural gas. Raw natural gas often contains several hydrocarbons. Pure natural gas (methane) exists as a gas, whereas many of the other hydrocarbons exist as liquids. Therefore wet natural gas is raw natural gas, before it has been purified.

2. Terms Related to Underground Water

Aquifer – An aquifer is a region of porous rock which contains groundwater. The water in the aquifer leads to wells and springs.

Ground water – Ground water is any subsurface water that is in the zone of saturation. Note that the top surface of the groundwater is the water table. Ground water is the primary water source for water wells.

Slickwater – Slickwater contains chemicals to reduce friction in the pipes. Slickwater is often used for fracking. Fracking slickwater will contain fracking chemicals, propants, and other additives, in addition to the chemicals added to reduce friction.

Watershed – A watershed is a region of land and associated drainage area which exists upstream from a major river.

3. Abbreviations Used in the Natural Gas Industry

- LDC = Local Distribution Company
- LFG = Land Fill Gas
- LNG = Liquefied Natural Gas; Note: LNG is not the same as NGL
- MAOP = MAxiumum Operating Pressure
- McF = Thousands of Cubic Feet
- MER = Most Efficient recovery Rate
- MMcF = Millions of Cubic Feet
- NGLs = Natural Gas Liquids; Note: NGLs are not the same as LNG
- PIG = Pipeline Inspection Gadget
- TcF = Trillions of Cubic Feet

Bibliography

Natural Gas

1. NaturalGas.org, www.naturalgas.org
2. American Gas Association (AGA) www.aga.org
3. Interstate Natural Gas Association of America (INGAA) www.ingaa.org
4. Office of Pipeline Safety (OPS), Dpt. of Transportation http://ops.dot.gov
5. National Pipeline Mapping System, http://199.107.71.24/publicsearch
6. Control of Pipeline Corrosion, by A.W. Peabody, 1978. National Association of Corrosion Engineers (NACE).

Chemistry Facts

1. The Elements, Second Edition, by John Emsley, 1990. Oxford University Press
2. Chemistry, by Wilbraham, Staley, Simpson, and Matta, 1987. Addison-Wesley
3. Web Elements www.webelements.com
4. Jefferson Lab Information about Elements
   http://education.jlab.org/itselemental/index.html
5. IEER Fact Sheets www.ieer.org/fctsheet/index.html
6. IUPAC (International Union of Pure and Applied Chemistry) www.iupac.org
7. IUPAC Compendium of Chemical Terminology
   www.iupac.org/publications/compendium/index.html
8. Lange's Handbook of Chemistry, 14th Edition, edited by John Dean, 1992. McGraw-Hill.
9. Table of Isotopes, Berkeley National Laboratory,
   http://ie.lbl.gov/education/isotopes.htm
10. Basic Inorganic Chemistry, 2nd Edition, by Cotton, Wilkinson, and Gaus, 1987. John Wiley and Sons.
11. Virtual Chembook, Elmhurst College,
    www.elmhurst.edu/~chm/vchembook/index.html
12. Modern Chemistry, various authors, 2004. Holt, Rinehart, & Winston.

Toxicology and Safety

1. Toxicology: The Basic Science of Poisons, 4th edition, edited by Cassarett, Doull, Ambur, and Klassen, 1993. McGraw-Hill.
2. Emergency Responder Training Manual for the Hazardous Materials Technician, by The Center for Labor Education and Research (CLER), edited by Lori Andrews, 1992. Publisher: Van Nostrand Reinhold.
3. Industrial Fire Hazards Handbook, National Fire Protection Association, 1979.
4. Toxics Release Inventory Program (TRI) www.epa.gov/tri
5. The Extension Toxicology Network (EXTOXNET) http://extoxnet.orst.edu
6. NIOSH Databases www.cdc.gov/niosh/database.html
7. NIOSH Pocket Guide to Chemical Hazards (NPG) www.cdc.gov/niosh/npg/npg.html
8. IPCS (International Programme on Chemical Safety), via NIOSH www.cdc.gov/niosh/ipcs/icstart.html
9. NIOSH Occupational Health Guidelines for Chemical Hazards http://www.cdc.gov/niosh/81-123.html
10. OSHA (Occupational Safety and Health Administration) www.osha.gov
11. OSHA Information for Toxic Metals www.osha.gov/SLTC/metalsheavy/index.html
12. ACGIH (American Conference of Governmental Industrial Hygienists) www.acgih.org
13. Office of Pipeline Safety (OPS), Dpt. of Transportation http://ops.dot.gov

Government Sites – General

1. US Department of Energy (DOE) www.energy.gov
2. US Department of the Interior www.doi.gov
3. US Department of Agriculture (USDA) www.usda.gov
4. Environmental Protection Agency (EPA) www.epa.gov
5. Food and Drug Administration (FDA) www.cfsan.fda.gov
6. National Institute for Occupational Safety and Health (NIOSH) www.cdc.gov/niosh
7. Mine Safety and Health Administration (MSHA) www.msha.gov
8. Federal Energy Regulatory Commission (FERC) www.ferc.gov
9. National Climatic Data Center (NCDC) www.ncdc.noaa.gov

Department of Energy (DOE) Related Sites

1. Department of Energy (DOE) www.energy.gov
2. Energy Information Administration (EIA) www.eia.doe.gov
3. [Office of] Efficiency and Renewable Energy (EERE) www.eere.energy.gov
4. Office of Fossil Energy (in Dept of Energy) www.fossil.energy.gov
5. Electric Transmission and Distribution Office www.electricity.doe.gov
6. Science (Office of Science) www.sc.doe.gov
7. Nuclear Regulatory Commission (NRC) www.nrc.gov
8. Civilian Radioactive Waste Management (OCRWM) www.ocrwm.doe.gov
9. Yucca Mountain Project www.ocrwm.doe.gov/ymp/about/index.shtml
10. International Nuclear Safety Program http://insp.pnl.gov
11. International Nuclear Safety Center, Argonne Laboratory www.insc.anl.gov
12. National Energy Technology Laboratory (NETL) www.netl.doe.gov
13. National Renewable Energy Laboratory (NREL) www.nrel.gov
14. Oak Ridge National Laboratory www.ornl.gov
15. Los Alamos National Laboratory (LANL) www.lanl.gov/worldview
16. Pacific Northwest National Laboratory (PNL) www.pnl.gov
17. Starlight, from PNNL/DOE http://starlight.pnl.gov

Index